The Observable Universe

Hannah Larrabee

LILY POETRY REVIEW BOOKS

https://lilypoetryreview.blog/

ISBN: 978-1-957755-30-4

Scan to hear Hannah Larrabee
reading from *The Observable Universe*

Contents

I am a student of nightfall, I claim no other profession.

—*Loren Eiseley*

551 Tons

Enter the poem as a plane through clouds, and you know
I read once that a single cumulous can weigh close to 551 tons
but we aren't afraid of them falling from the sky. They sit
around basking in sunlight, admiring the rolled bales of hay
below. We don't see them form, we don't see them dissipate,
but we do see them come together, occasionally, when patience
allows. And then sometimes we transcend them and the flight
evens out at 30,000 ft and there's a little ding and we unfasten
our seatbelts and it's as if we are weightless just simply
levitating above an endless mattress of clouds.
There was an inscription at Persepolis that read
Whatever seems beautiful we built by the grace of God
I don't think they had only human hands in mind
though it might benefit us to build as if planning for
future ruin, what it might look like under the weight of time.
At the Parthenon, I stood in the shadows and looked up
at the great columns and really the only thing that moved me
was a sliver of light, how gentle it was on the stone.
There is a temple we each tend like Lhasa Apsos;
imagine, for a minute, the polished stone. Interiority,
I am always working on interiority, and silence accumulates
in places where once there was not music exactly, not even
conversation, but a humming that meant I was alive
without ever having to think about it. But I think about it.
And once in a while, I have the kind of a dream
that takes me for who I am and I can't believe
I can never just say what I want to say.

Deep Adaptation

my body   a chrysalis   my jaw

steel shut   too busy becoming   that which the earth

has spun for me: a mud hollow   barn swallows

great sadness   of the biosphere   falling silent

why didn't we   unbuild our cities   stone by stone

I still dream   of kissing you   but I also listen

to the whispers   of trees   the chatter of birds

what they are saying   *about us*   and there is

what feels like   a light left on   behind my eyes

live circuitry   of a language   so afraid

it is the same word   over   and over

Stress Eating Cheetos

It could be the constant gun deaths or maybe
the politicians high-stepping through the Capitol
rotunda at the first sign of the violence they provoke,
or the extinction of Bachman's warbler, or my worry
about the long-necked Bird of Paradise, or Ron DeSantis,
take your pick, dear Cheetos, I open your bag and
it's hard to stop, *why stop*, even if you could answer
that question this is all by design, evolving to the point
of Cheetos and once Cheetos then an ever-perfected Cheetos,
flavor to texture to tongue a golden ratio, a fractal repeating
in each Cheeto, and maybe you already agree with me
so the question isn't *is this good for me?* It's *why have I chosen
the vice that leaves all the evidence?* My fingers stained electric
orange, even after I've moved from the utility of fingers
to my more generous mouth, as I would with a lover,
I'm thinking this color isn't anywhere else in the multiverse,
and even if neon orange dusts the surface of some distant,
lifeless planet, I'm sure Cheetos only appear in *this* universe,
the Cheeto-verse, as one might call it, and there is a question
I wish every gunman would ask himself, of all the harm
imaginable, why can't it just be this: the growling disapproval
of a stomach opposed to preservatives, not the shattering
of bodies, the eviction of so many beautiful souls, just once
let a friend fall asleep with their head on your shoulder,
it takes a softer version of the world you build inside, Cheetos
are just one language for the difficult now that must be fed,
and as an act of resistance, I imagine Cheetos stuffed into
the barrels of guns, or an entire bag tossed to the floor the way
she dropped my jacket so we could hold each other because
the world blessed us with this—the closer orbit of our lives.

To Render a Deer as Your Life

Three Christmas trees,
now Douglas firs in their own right
20 feet tall and never harvested
because my father planted them
and then my father left,
they are sentinels of the same scene,
same sunrise, same kaleidoscopic
stars, and there is my mother
looking out the kitchen window
in the morning with her coffee,
but this time there is something
under the trees, a bulk, a stillness,
some faint outline of life. I wasn't there
but it felt like I was when my mother spoke
of the doe curled up "like a dog," she said
and I knew what she meant, which dog.
It turned her stomach, tracking drops
of blood up the gravel driveway
to the road, or maybe it was the idea
of the deer seeking shelter beneath
these trees, on this land, the desire for it
simple and strong as water pushing up
from ancient aquifers. Something moves us
in an unnamed direction; death, always
death with its dull brown eyes.
In the branches of those trees
were generations of mourning doves.
They had tucked their beaks into
the warmth of their own feathers,
assumed the same shape as the deer
—*shape of what*? I couldn't say.
I once stood watching the red blinking
of Mars and heard the movement of a deer,
its hooves at the edge of the driveway.

In the darkness we stood breathing
and there was something that could see
all around us and it did not want to speak.

Night Market of the Ghosts

One day, when your spirit is exhausted, disasters will arrive.
—Werner Herzog's *Into the Inferno*

It's true then: something this way evolves
with invisible teeth, all our rigorous and creative
confrontations, meaningless. Nothing changes
all at once, but it changes completely and to us
there is no difference in mourning. Free yourself
enough to ask, *am I different*? If I see the same
things day after day, *am I different*? I think
I went barefoot in the grass for the first time
in a long time. I made some changes to the way
I regretted not saying things how I felt them,
or not recognizing the way she led me to things,
I was so caught up in what would happen.
The thing of it is, all around us, it exists
or doesn't exist and we can't tell until it strikes
us close enough to hear it breathing. I don't want
an obscure number of people to die, for any reason,
virus or violence. I only sometimes feel brave
enough to think about how many insects exist
at once in the valley fields. Maybe there is something
to disaster as *recognition*, of spirits taking up
their wares in the night markets. It makes sense
to worship a volcano, to withdraw into our own
homes at the earliest rumble. That we must
quarantine is new to us, and I guess I thought
the earth would uproot us the way radishes
are taken in spring, but I never envisioned
the mouth we'd feed, and it turns out there isn't
just one mouth there are many and so few things
to say about them that I feel I have been sent
back to myself, like the teenager I used to be:
up late, alone, drawing at my quiet desk.

Smokey Bear on Tinder

38.

Purveyor of
green

and alchemist
of fire's

cousin, which is
not-fire.

Interested in:

Netflix and
BEWARE

the unintentional
spark

Netflix and
CARRY IN,

CARRY OUT.
Things I enjoy:

the absence
of wind,

a lover
with a body

of rain,
or a body

of sand, please
show

with wildflowers.

I've never
once

closed
my eyes

not even to kiss,

think of all
I've given up

to stop what
happens anyway.

A few things you
should know:

I might speak
softly

until
it is necessary

to yell,

if you've
loved a forest

then you've broken
with fire

yet dream
of it, *dream of it*

devouring
the underbrush,

please show
your animal

face.

Earthshine

Look at the moon, not too many days after new moon, and you'll notice that the dark part is also visible because the Earth is shining on that part of the lunar surface. It's Earth light.

—Maggie Turnbull, Astrobiologist

Let's be honest, it's a mess of light
and the moon sometimes likes to pull
the shades. I think of the moon like
the bucket of ash after several fires,
if the bucket had a pulse. Sometimes
embers make it hot to the touch,
but the moon sleeps without covers
in the perpetual winter of space.
We barely see winter on Earth now
and my skis can't take me anywhere;
if only the ashen powder of the moon
let me glide into its craters. I think,
how erotic, long nights spent gazing
at the moon from Earth, to lie back
and let it pull the oceans. If I were
on the moon, I'd talk to myself a lot
not to feel like my voice might carry
out into the vacuum, but under my feet
I think the vibrations might mean
something. You can't convince me
that this yard is the only holy ground
for moles. Some say the best soils pass
over their velvet bodies. I think,
if the moon had moles, they might
regret the work of their lives without
the vibrant humming of the Earth.
Earth moles have very limited sight
but know, intimately, the blooming
all around them. And if something

catastrophic happened to the sun,
on the moon, at the right point of orbit,
I'd know in 6 minutes leaving 2 minutes
until the Earth understood. And the moles,
monks inside their silent tunnels,
could never make the mistake of assuming
this was all for them.

One Theory

If the same dandelion sprouts again in the same field,
the honeybee might recognize its sweetness but we'll
never know. The firefly flashing in front of your face
is the same one you see now in the plum tree. I speak
of earthly things with such a fierce nostalgia; I replay
the losses, memories: stubble of pumpkin vines,
my father's propensity for silence. I was not ready
for such memories when I used to touch everything.
Now touch opens great fissures in me, one person
kissing me, touching me, intricate as tangled roots. I have
been careful not to take anyone to bed who I do not want
to remain in some way. I don't think I will ever be okay
with the temporary lending of atoms, the porous and
unstructured body of desire. Just once, I'd like to rest
with my full weight dissolving into the bed, the grass.
Are you interested in one theory: one body, one death,
one saturating feeling? I never considered how many
frogs sing from vernal pools at the ache of spring; to me
it is one song. And I have never needed my body to come
back to me more than I do now. It could walk from the fields
around the corner of the barn straight into my arms.

Machine

Say we try this again   say we know it best with our eyes
closed, oiled fingertips   the thing we want is machine
is mechanized tongue; in these words, in their primary shape
I betray the evergreens;   nothing can stop the industry of
shame, slick with design   and no less deadly to the soul
that wanders in nothing but burlap   singing to the trees;
I was once this soul   and now I want to be tied to a bed;
who best to serve me: man, woman, machine
control is every gear spun out   of the minerals of the earth
that have forgotten their time underground   I am
on a train taking me to a concrete city;   skyscrapers feign
sky with glass   how else to reflect what was good of you
in me   I must forgive this poem   every poem   each one
betraying you to me in parts   and not the whole, elegant
machine.

The Observable Universe

We've all heard the question *why is the sky blue* but *why is it dark at night* is something else entirely. It becomes a story beneath the science, like infrared of Botticelli's *The Man of Sorrows* revealing a sketch of Madonna and child. What did he decide? Was it prayer rearranged into another prayer? We are left with what is observable, what we are meant to see. The universe itself is illuminated just right; what we see is light having traveled this far and that's it. So much light is still on its way. We'll never see it all, no human being, but maybe a tarsier with big soft eyes or some distant relative gripping a tree, the night sky getting brighter. I'd hate it. Leave me inside the sensuous dark or at least be bold enough to bring me a brighter lover to obscure the sky: Andromeda, or maybe Saturn and its rings taking a closer seat, the scale model of desire collapsing. 12 degrees tonight and a ladybug clings to the kitchen light. The stars are clear as hell in this crisp air. When I ask *what is observable* what I mean is *what is still making its way to me?* I've been told we create the kinds of relationships we want, and of course that's true but it doesn't consider the distance required to travel. It matters sometimes to be far away. I am standing watching the ladybug wondering which prayer was closer to Botticelli —the one he hid from us, that seems right.

Rain Frogs

If you can just imagine
a migration among the night fields—
several inches of rain the weight of it
suspended, for a time, among the blades
of tall grasses, every land a wetland
so the frogs embrace their full dominion
and they leap across pavement because
pavement is just a shallow dark pool—
a portal to the other side where there is
another side, I don't know—the mind
of a frog is music, is a full-throated song—
I have felt one breathing with its whole body
in my hands, as a child they were tadpoles,
they were eyes with tails kept in pondwater
in a painters bucket
in the primordial beginnings of life
like gelatinous bubble wrap if nature
taught us something of the softer elements
of the earth—can you imagine the love of it,
of the earth, placed inside your hands
like a wet stone? I am a child again
and always beyond saving if saving
is not this—saying goodbye to myself
in a torrential May rain—
barely a moon if there was a moon
and among the grasses a tectonic shift,
colonies in motion, little green whales—
acres meant for somewhere or something
unlike me, not in that way, not until
I understand that what I hold in my hands,
what I watched watching me in those pale buckets,
was never mine to have or to lose but to tend,
to tend, no different than curling into yourself
to listen closely like the heart is a cave
and it echoes a sonic and drenched kind of
love for you, God—

Coyotes

If the stars are the hunters we imagine them to be,
then they are coyotes

short radio yips, pulsars, eyes of honey—

and they will pierce you as they did me the time
I turned around and there they were

standing, hovering, their necks moving closer to the ground—

so as to see me in the way that cut, reduced me
to a pocket life, a want for wild

and upon arrival a longing for, I can't say—

Information is Matter and the Heart is Matter and I'm Afraid the Universe Pulls it All Apart

But sure, PSR B1257 +12b
spinning 2,300 light years away
could be a veritable picture
of Earth-like conditions, imagine
the soft trade winds of Aruba
mixed with a slight blue haze
that is, if 12b wasn't orbiting
a pulsar blinking on and off
like fluorescent lights in public
school hallways, drama leaflets
bursting with radiation

In—
most—
things—
could it—
be this—
simple—
I mean—
lifeless—

But the Webb telescope
is just now lifting its shades
to see, bending across
some invisible arc,
the red embers of information
and some might ask, *where is the heart*
in all this, where is the heart
in your reading? I think,
under the black tongue of space,
we might be nothing more

than the form information has taken
and aren't we
and haven't I
peeled the backing off my life,
deployed a five-layer sunshield,
recalibrated, and now remain still enough
to gather light
but when I said aloud
I am ready for softness
softness never came ...

Is it the same out there
in the universe, all our chairs
facing out in an endless room
of empty chairs, and if not
will we finally have something to say,
or will it be more like the time
I stayed up all night with a glass of water
beside the bed and dared not drink it,
the envelope of the night too delicate,
the delicate assembly of love,
but I was wrong, love was there
without my invitation, it assembled
even when others said there was not enough
information, and didn't it reach you,
somehow, whenever it was you noticed
that glass of water beside the bed.

Birdsong

We have a very discreet bandwidth of super-sensitive hearing between 2.5 and 5 kilohertz… is there something that matches our peak hearing human sensitivity? Indeed, there's a perfect match: birdsong.

—*Gordon Hempton*

I want to live here, I always have,
but I was so often alone, playing in attics
or clinging to the legs of trees. I grew,
and began to endanger myself in the ways
I had been endangered, and there was never
a conversation about this, or if there was
it was below the frequency of birdsong
so I did not hear it. Now, I want to destroy
all presumption that we have evolved toward
the human voice when we could have adapted
to birdsong. I wonder, is there a measure
of softness like frequency, a kilohertz
of softness? I would listen differently here,
aware of the change. But, as it turns out,
we just couldn't flourish in abundance.
I don't know how else to say this …
sometimes a lone cardinal remains at dusk,
and sometimes I know he's there before looking.
The clouds are so low here; I want to know
when they will touch me.

Giant Groundsels

Dendrosenecio is a genus of flowering plants family native only to the altitudes of Mount Kilimanjaro. This poem is in memory of Oliver Sacks, especially his work in *Oaxaca Journal*.

Cowboys of Kilimanjaro,
wearing chaps, rooted in dust,
and rare: preferring one mountain
more than any other,
rare in sensibility, too,
and standing on impossibly
small legs, legs used to spending
half a life hovering
on the back of a tamed beast,
if you consider tame
accommodating to a hilt, a halter,
a whip. I think the giant groundsels
have no idea how outlandish they are,
how totally replaced,
over a million years,
they have become;
they do not care. They open
toward what feeds them;
they sculpt themselves
awkwardly in prayer
to the mist, to the elevation,
to the absolute freedom from us
most of the time,
and if they sleep they dream
of things more like them, imagine that:
ancient dreams.

Indigo

In memory of Oliver Sacks

At the Metropolitan Museum of Art,
I once stood transfixed at the fragment
of an Egyptian queen's face, falling in love
with a color, yellow jasper, and her lips.
Whatever it was you could not quite reach
—there you spent your hours. I think of it
when you said Giotto could never paint
the holy skies indigo—he was not aware
the color of heaven no longer existed on earth.
What have we done with all these gifts?
The proud chest of a grackle flashing in the sun
—not indigo, but indigo as it can be seen by us
shrouded in black, or left behind in the teeth
of medieval scribes. Now that we are clean,
technologically clean, what will they find
in our graves? No seeds, no weapons, no vessels
full of mead, but devotion enough can become
a kind of madness ringing through millennia.

Paintings of Interiors

I never went down my grandmother's spiral staircase when I was young.
The house was already haunted, a séance held every year behind closed
doors. Is it enough to be afraid, because I am afraid when I think of
Saturn's moons locked in a gaze with something not really there. I know
Saturn is there, maybe even more so than Neptune with its fledgling
rings, or else where did Cassini take its final plunge at 6:32am on
September 15th 2017? I get a little scared contemplating all the churning
business of space. I tend to think of it when I walk through doorways.
Some people feel this way when they step inside a church. Inside a church,
I only want the paintings, the stained-glass depictions of Christ. Portals,
I think, to the many faces of fear, like old heating grates in the floor. If
you've ever listened to the sounds of the planets then you know they're
haunting. As for Earth, I wonder what it sounds like from far away. There
is warmth when I think of it. I remember there was one childhood night
when I tiptoed out of the room to the staircase. I tended to step directly
into the things I feared, but that ended when the ocean changed its tone
and I was afraid to go in. If you are looking for the horror we've brought
upon the Earth, it's right there. This somehow includes the oceans of
Saturn's moons. Physicists describe it as *interactions* that no properties
exist outside *relations*. Ghosts certainly aren't the best example. Maybe
the best example makes no sense to us, like digging up tropical fossils
in the Arctic. What we see is not what has been and it breaks us—the
entire briefness of us. The trick is how to carry it for so long, this interior
painted all around me. And just outside the window: Saturn's moons.

Artemisia

I have taken the morning
to read about myself by reading
about Artemisia Gentleschi,
and the graveyards of airplanes
in deserts, airplanes that still
try to take off with any semblance
of wind, and could Artemisia
have sold herself on talent alone,
yes absolutely yes but then talent
is constrained in order to be
whatever it is that turns the pages
of history, of commerce, a woman
is banished in her own way, it is
unique to each of us, and I think
of her paintings in museum stillness
separated as they have been from
her touch, four hundred years
to think themselves over in the eyes
of strangers remarking about
the boldness of color, the weight
of shadow, and Artemisia never saw
an airplane, at least not in the air,
and since I have seen both:
her paintings, the airplanes,
does that make me someone
who could explain—anything?
No; you see, when I take something
in for the first time, I am an ancient
single-celled thing, lust of senses,
something of color or form, lapis lazuli,
or second wing, yes, I am emboldened
to learn more about myself in these
unrelated things, but as Artemisia said
never has anyone found in my paintings

a repetition of invention, not even a single hand
and in finding nothing else to relate
these things to, I turn to memory,
which isn't a thing at all but like
a painting, witnesses me where I stand
and I lift a little from my shoes
on that air, on that body

Say Something

Why spend all this time in the byways, the perennial streams,

when we are already in relentless communication,

if thought is communication—if intention, too, is communication.

I know what is happening and it weakens my heart, its structural integrity.

Now I take pills for it, even though I don't want to endure

the intimacy of knowing but not saying, or not knowing what to say

but wanting to say, at least, something—to kickstart whatever it is,

bring it outside to warm in the sun. But that's the thing—

I am always the one stepping off the platform onto train after train.

I swear, if you put a single word in my hand first

I would become an oak shedding its leaves, a tanager at rest.

Home is a Rogue Planet Careening through Interstellar Space

In some dark corner of some
elaborate library of the cosmos,
there is a book loosely held together,
more a suggestion of a book
but a book nonetheless,
where it is written that I will not feel
home anywhere in this life, though the search
continues as the soul, carved through time
and God's attention, has its own memories
of a time when home was a wheat field
or an outcropping of rock
or maybe it was not on Earth at all.
And what this has done to me
is make me love the infinite: space, yes,
but also the well-trodden deer path through
a vernal wetland near the Saco River
where I camped once and was awoken
suddenly by something hitting my tent
then car so hard the antenna was still rattling
by the time I unzipped and looked out
with my small flashlight. Once,
I sat with my feet over the side of the bed
and put my head in my hands and prayed
for whatever was meant for me to come.
And the rest of the time I have been trying
to find it by loving each thing I encounter
on Earth so much that the final flicker of the tail
of a red squirrel hit by the car in front of me
feels like someone scraping the last of something
from the container of my heart. And that's just it
really: why contain it. Even at the level of atoms
we are barely held. What beckons when we look up
at the stars is that someday we have to leave.

Or be ourselves only in transfiguration. Like God,
home is a word that knows when you mean it.
Someday it will dissolve upon my tongue.

Ice-Albedo

For some people the day comes
when they have to declare the great *Yes*
or the great *No.* […] Yet that no—
the right no—drags him down all his life.　—C.P. Cavafy

Save the lesson in thermodynamics, these days
even the average afternoon light seems harsh.

Can you relate, Cavafy, from whatever expanse
you are precariously contained?

> I say, if the moon
> comes close enough
> to lick, by all means:
> salt the tongue, drag it
> through the treeless
> and blazing dirt …

So far, the word *lush* exists in two places:

> on Earth, in the corners without us,
> and at the turn of one kiss into another

Cavafy, I am pleased to announce
that you have been absolved of any climate complacency

even the gluttony of bath houses where you said a *Yes*
that sounded like *No* …

> (and why should we even care about the reflective properties of ice
> when we are hot as the jaundiced eye of a Yellowstone spring?)

I never thought a lover could say the opposite of what they want
and, truthfully, ice is rarely considered for its quality of light

even as it deflects it
from Earth

in what we might perceive
as an act of great
mercy

saving us from the dregs, the thawing permafrost,
the starving grasses, saving us from saving us from
saving us

I Can't Explain It

I felt his departure
so clearly even his last breath
passed through my hands
it was his thin body
it was his own body
that kept him here, after all,
and then he didn't come
to say goodnight
to the right of my pillow.
And in the morning,
though I checked, his little bed
was empty. I have almost nothing
to say about grief, its tongue
tucked under my heart.
I had to say his name
for him to lift his head,
now his name is everywhere
and in nothing I can touch.
All day I feel it in my body.
And at night, standing beside
the July fields, the July trees,
all the electricity of the crickets,
I can't explain it
but you are listening to the stars.

Acknowledgements

My gratitude to the editors of the following journals where these poems first appeared, sometimes in slightly different forms:

"Stress Eating Cheetos" in *Action, Spectacle*

"Artemisia" in *The Ekphrastic Review*

"Birdsong" in *EcoTheo Collective*

"Machine" in *The Maine Review*

"Smokey Bear on Tinder" and "Coyotes" in *Gertrude Press*

"One Theory" and "Night Market of the Ghosts" in *Voices Amidst the Virus* by *Lily Poetry Review*

"The Observable Universe" in *Flypaper Lit*

About the author

Hannah Larrabee's *Wonder Tissue* won the Airlie Press Poetry Prize and was shortlisted for a Massachusetts Book Award. She has a recent chapbook of poems, *Dear Teilhard,* exploring climate change and spirituality out from Nixes Mate Press. Hannah's had work appear in *Flypaper Lit, River Heron Review, Gertrude Press, The Maine Review, The Adirondack Review, Glass: A Journal of Poetry,* and elsewhere. Hannah wrote poetry for the NASA James Webb Space Telescope program and read her work at Goddard Space Center. She participated in an Arctic Circle Residency with artists and scientists in October 2022. Hannah received an MFA from the University of New Hampshire where she studied with Charles Simic. She taught writing at Northern Essex Community College and The New Hampshire Institute of Art/New England College. (pronouns: she/they)

Printed in the USA
CPSIA information can be obtained
at www.ICGtesting.com
JSHW020219011123
50975JS00004B/67